Gohar Marikyan, Ph.D.

ANANIA SHIRAKATSI'S

TVABANUTIUN: WORLD'S OLDEST

MANUSCRIPT ON ARITHMETIC

PART 1: ADDITION

2nd Edition

$x + y = ?$

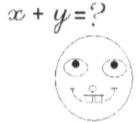

New York, U.S.A.
2019

1

Anania Shirakatsi's Tvabanutiun: World's Oldest Manuscript
On Arithmetic
Part 1: Addition

by Gohar Marikyan, Ph.D.
2nd Edition

Anania Shirakatsi's 7th century manuscript *Tvabanutiun* (Arithmetic) is the world's oldest preserved manuscript on arithmetic. It contains interesting detailed explanations of the methods he has developed for teaching arithmetic to beginners and to first grade children. Shirakatsi's methods have been successfully used over centuries in Armenia. Even though times have changed, his methodology continues to be as effective in the contemporary diverse classroom environment as it has been since the 7th century.

New York, 2019

ISBN: 9781085940726

To my father
Gerasim Marikyan, Ph.D.
who was my inspiration.

"I love Armenian people - all of them. I love them because they are a part of the enormous human race, which of course I find simultaneously beautiful and vulnerable."

William Saroyan

Content

Preface 7
1. Historical Background 9
   1.1. Introduction 9
   1.2. Anania Shirakatsi's Manuscript 13
   1.3. Numbering Systems 24
2. Anania Shirakatsi's Methodology 30
   2.1. Net Effect of Anania Shirakatsi's Methodology 30
   2.2. Anania Shirakatsi's Tables of Addition 31
     2.2.1. First Column: Addition of Single Digit Numbers
     33
     2.2.2. Second Column: Addition of Tens 36
     2.2.3. Third and Fourth Columns: Addition of Hundreds and Thousands 37
3. Conclusion 39
ABOUT THE AUTHOR 44
OTHER PUBLICATIONS 47
Index 49
References 53

Preface

Anania Shirakatsi, in his 7th century manuscript *Tvabanutiun* (Arithmetic) describes the methods he had developed for teaching arithmetic to beginners and to first grade children. This manuscript is the world's oldest preserved complete manuscript on teaching arithmetic (Hewsen 1968, pg.42). It contains interesting detailed explanations. Shirakatsi's teaching methods have been successfully used over centuries in Armenia. The net effect has been the acquisition of higher levels of arithmetic knowledge by children, setting a strong foundation for achievements in scientific research in their adult years. Even though times have changed, his methodology is as effective in the contemporary diverse classroom environment as it has been since the 7th century.

This book is only a synopsis and a brief introduction to Anania Shirakatsi's genius and his methodology in teaching arithmetic. There is much more to research and to talk about this fascinating topic and about the beauty of his work. The aim of this book is to reflect on Shirakatsi's methodology for teaching addition to beginners and to children.

1. Historical Background

1.1. Introduction

Mathematics, as a science, is one of the pillars of civilization. Even though history may ascribe its birth to a particular nation, the fact is that mathematics, most probably, started with Adam and Eve, and most probably its origin is connected to the ten fingers that we have. Perhaps that is the reason why we use the ten-based numbering system albeit there have been civilizations that have used numbers in other bases. When it comes to teaching mathematics every nation has its own unique method that has been developed over their own history. A few of these methods are interesting indeed. The study of history of mathematics helps the contemporary educator determine the necessary tools for teaching mathematics to today's students. A number of old manuscripts contain a wealth of information that shed light to the historic development of mathematics, nurturing the continuation of current thinking.

When it comes to major contributions, one such nation is Armenia, and one particular person in that nation is the 7th century mathematician Anania Shirakatsi, the author of the world's oldest preserved manuscript on arithmetic.

In this book I will review and reflect on Anania Shirakatsi's fascinating methodology and his ingenious approach to teaching arithmetic to beginners and to children.

Although Anania Shirakatsi has left a vast array of scientific works, their discussion is outside the scope of this book. I will briefly discuss the history of the manuscript and

**Picture 1. Statue of Anania Shirakatsi at Matenadaran,
Yerevan, Armenia**

the biography of its author. Additional information about

this seventh century Armenian mathematician, writer,

11

historian, geographer, cosmologist, and philosopher can be found in Hewsen's article (Hewsen 1968) which includes excellent footnotes and references, Abrahamyan's book (Abrahamyan 1944), Hacikyan's book (Hacikyan 2002), and others.

It should be noted that quite often people use the word mathematics instead of the word arithmetic even though the latter is only one part of the former. The word mathematics is from the Greek word mathēma, learning, while arithmetic is from Greek words arithmein, to count and arithmos, number[1]. Even word analysis shows that mathematics is much inclusive than arithmetic. Mathematics is a theory, while arithmetic, being its small part, deals with addition, subtraction, multiplication, and division of numbers.

In this book I refer to very basic arithmetic, that is, the very

---

[1] Merriam Webster's Collegiate Dictionary, 12th Edition

fundamental art of adding numbers.

**Picture 2: Statue of Mesrop Mashtots with the Armenian
alphabet, at Matenadaran, Yerevan, Armenia**

1.2. Anania Shirakatsi's Manuscript

At the end of the 4th Century the necessity emerged to
revive the lost Armenian alphabet. King Vramshapouh and
Catholicos Sahak Partev assigned the task to Mesrop

Mashtots, a genius scholar monk. In 405, Saint Mesrop (see **Picture 2**), revived the alphabet of the Armenian language, consisting of 36 letters. To date it remains unchanged, and it is currently in use by Armenians (later, three more letters were added). Soon after Mashtots's 36 letters were put to use in translating the Bible. Scientists and scribes were inspired to write books in various fields of science, history, literature and religion, and were encouraged to translate scientific and religious works from other languages. For the next two centuries, political unrest paralleled the exceptional development of scientific, literary and religious life that became known as the first Golden Age of Armenian literature (5-7th centuries). The 36 letters of the alphabet also replaced the previously used numbering system, and it was Armenia's official numbering system over the next 1200 years. It has survived the adoption of the internationally used Hindu numerals[2] and it continues to be used in Armenian religious literature and in other writings.

My hero is Anania Shirakatsi, the 7th century Armenian mathematician (see **Picture 1**). His manuscripts have been of great interest to scholars over centuries. Grigor Magistros

Picture 3: Anania Shirakatsi's manuscript displayed at the Matenadaran in Yerevan, Armenia

---

[2] Are also known as Hindu–Arabic or Arabic numerals.

(990-1058), an Armenian linguist, scholar and public functionary, among other manuscripts of scientific and philosophical value also owned several works of Anania Shirakatsi. He states that Shirakatsi's works were of great interest to scholars of the time[3]. Today, there remains only a small portion of Shirakatsi's manuscripts that has been handed down to us. They are currently among a very vast collection of manuscripts safeguarded at the Matenadaran (see **Picture 4**), Armenia's Conservation and Research Center of Ancient Manuscripts and Documents. The Matenadaran is in the capital of Armenia, Yerevan. It contains some 17,000 irreplaceable manuscripts in several ancient languages (Armenian, Greek, Latin, Arabic, Farsi/Iranian, Assyrian, Hebrew, Ethiopian, Indian, etc.), and in total more than 300,000 documents, thus, the largest collection in the world. **Picture 3** is a photo of Shirakatsi's Tvabanutiun manuscript displayed at the Matenadaran.

---

[3] http://www.nationmaster.com/encyclopedia/Grigor-Magistros

Picture 4: Matenadaran, Yerevan, Armenia

Shirakatsi's survived manuscripts are: *Inqnakensagrutiun* (Autobiography), *Tvabanutiun* (Arithmetic), *Tiezeragitutiun* (Cosmology), *Tomaragitutiun* (Chronicon[4]), *Ashkharhatsuyts* (Geography), *Odyerevuitabanutiun* (Meteorology), Chapagitutiun (Weights and Measures).

Anania Shirakatsi is one of a few Armenian authors that left a detailed autobiography. Similar to other natural sciences

---

[4] Comparative chronology of celestial events

related original sources, Shirakatsi's writings and autobiography tell us about the degree of development of science at that time and also gives an idea of the art and means of constructing in that era, and the level of production in general. This data supplements the shortage of historians' data. Shirakatsi's autobiography also contains facts about his life.

Anania's year of birth is generally known to be between 595 and 600 A.D. Shirakatsi's autobiography shows that he was born in Ani borough of Shirak province in Historic Armenia. That is why, as was the custom in the olden days, his name simply means Anania from Shirak. Shirakatsi received his primary education at a local monastic school, where he developed great desire and passion for knowledge.

Shirakatsi was a well-educated young man. He studied Armenian literature and the Holly Scriptures, and was highly impressed by references to wisdom in the Holy Bible ("The wise shall inherit glory, but shame shall be the legacy of fools." Prov. 3:35, and the like). In his own words, "With great admiration of the art of calculation, I came to the conclusion that without numbers nothing can be reasoned, therefore, I considered it the mother of all knowledge." (Shirakatsi, Autobiography, 7th century)

Sirakatsi's approach and the love for mathematics and sciences has been passed down through generations and I, a child of the 20th century was a recipient of Shirakatsi's message, and I was taught to love and to respect mathematics, and knowledge (education) in general. Isn't that amazing, to be a recipient of a 7th century message?

In search of a teacher, on the recommendation of his aquaintances, Anania went to Trebizond to meet Tychikos, a famous scholar and professor[5], who accepted him as his pupil.

---

[5] Rosenqvist 2005, p. 33

Shirakatsi studied under Tychikos for eight years. After returning home in 651 Anania immediately founded his own school, although at the time, there were a number of high level schools in Armenia (Syunats School, Dprevanq School in Shirak, Arshakunyats School in Yeraskhadzor, Mairivank School in Bjni, Glak School in Taron, etc.). Shirakatsi's school most probably was to prepare students to continue their education in those schools.

The 7th century manuscript *Tvabanutiun* (Arithmetic) is Anania Shirakatsi's textbook that he used for teaching arithmetic. In his manuscript he describes methods he developed and used in teaching arithmetic to beginners and to first grade children.

[Maintaining its commanding aura, as I can best translate from the Armenian text,] Shirakatsi writes, "You, who seek wisdom and wish to study with me, as an alive voice of a good teacher, my purpose is to write for you the science of computing that has been developed by the efforts of our ancestors. Learn the tables that I have formulated. Although I have abridged the existing comprehensive version to avoid boring you with lengthy repetitions, I have also simplified the original concept so that you understand it profoundly and thoroughly. So, I start with the most fundamental and the simplest, having in mind the learning level of children and the uneducated."[6]

Shirakatsi's emphasis being on children and the novice, he avoids teaching by rote and prefers teaching by understanding. If the child understands why and how two numbers add together, she/he does not have to memorize anything, any step, any method, any formula, or any algorithm. This "understanding" begins with the most fundamental of the four operations of arithmetic, the addition.

---

[6] Aghayan 1979, p.30

"Nakhavarjzoom, the first chapter of arithmetic is called Addition."[7]

By "the most fundamental and the simplest" Anania Shirakatsi refers to the operation of addition, and he refers to its process of learning as *nakhavarjzoom* (preparatory). A good understanding of addition makes a strong foundation for learning the other three operations and facilitates the learning process in acquiring further knowledge.

The appendix of this manuscript contains several "Khrakhchanakanner" (Problems of Amusement) that Shirakatsi had meant for mathematical entertainment in social gatherings. Scholars that have studied this manuscript are of the opinion that perhaps it contained a theoretical part that has not survived.

---

[7] Aghayan 1979, p.30

From the existance of mathematical "**Problems of Amusement**" in Armenia in the 7th century we can conclude that then mathematics was loved, well understood and pretty popular among people in Armenia, because logical thinking then would give people enjoyment. Therefore, mathematical problems had become a part of entertainment in gatherings. I find this fascinating, especially comparing it with the nature, quality and level of current entertainment, which does not contain any element of thinking. This reminded the phrase I have heard from my students, "Please, don't make me think."

"Shirakatsi's spreading of mathematical knowledge in Armenia is especially significant if one realizes that the period between 500 and 800 A.D. is generally considered the dark ages of mathematics in the West. The Greek Proclus (c. 410-85) and the Roman Boethius (c. 480-524) were the last glowing embers of the dying fire of mathematics; in fact, the historian Edward Gibbon characterizes them as 'the last of the

Romans whom Cato or Tully could have acknowledged for their countrymen.' (Quoted by D. E. Smith in History of Mathematics, New York, 1958) Some three hundred years were to pass before mathematics was revived by Arabs, in the ninth century."[8]

1.3. Numbering Systems

In ancient times, before Hindu numerals were invented, Greeks, Armenians, and some other nations used letters for numbers by assigning values to the letters of the alphabet. This system is called Ionic Numbering System. Just after revival of the Armenian alphabet in 405 A.D. by Mesrop Mashtots, the 36 letters replaced the previously used numbering system in Armenia. For that purpose, Mashtots arranged his letters of the alphabet in four columns with nine letters in each column; each column progressing in a vertical fashion top to bottom,

---

[8] Hacikyan 2002, p. 57

and then continuing to the next column to the right. (Most depictions of the Armenian alphabet are in this fashion (See **Pictures 2** and **5**).

The first column represents ones, the second – tens, then comes hundreds, and the fourth column represents thousands. Therefore, each column represents one place value. There was no letter or notation for zero in the Ionic Numbering System is system because it is not a positional numbering system, and there is no need for zero. As a result, numbers in this system are represented in more efficient way. For example, in **Picture 5** we can see that the number 3000 is represented by one letter, Վ, which is the third letter of the column for thousands.

To write a number we need to use letters, so that the sum of their numeric value be equal to the number. The letter with a higher place value is placed to the left of those with lower place values. For example, to write 2009 we need to write the

Ones		Tens		Hundreds		Thousands	
Letter	Numeric Value	Letter	Numeric Value	Letter	Numeric Value	Letter	Numeric Value
Ա	1	ժ	10	ճ	100	ո	1000
Բ	2	ի	20	մ	200	ս	2000
Գ	3	լ	30	յ	300	վ	3000
Դ	4	խ	40	ն	400	տ	4000
Ե	5	ծ	50	շ	500	ր	5000
Զ	6	կ	60	ո	600	ց	6000
Է	7	հ	70	չ	700	ւ	7000
Ը	8	ձ	80	պ	800	փ	8000
Թ	9	ղ	90	ջ	900	ք	9000

Picture 5: Armenian alphabet with corresponding numeric values

letter with numeric value two thousand followed by the letter for nine (see **Picture 5**). Below is 2009 in Armenian alphabetical numerals.

Ս Թ

For 9999 we need to write the last, 9th letters of columns for thousands, hundreds, tens and ones (see **Picture 5**). Below is 9999 in Armenian alphabetical numerals.

ԹՂՉ Թ

Again, by adding all the numeric values of the included letters we will get 9999.

9999 is the largest number that can be written using the 36 letters of the Armenian alphabet. An additional sign, the sign *byur* similar to ^ was used in conjunction with letters to write numbers higher than 9999. It was placed above a letter to indicate that the numerical value of the letter is multiplied by 10,000 (see **Table A**).

The principle behind this system is similar to the Ancient Greek numeral system. However, there were major differences

Table A: Shirakatsi's tables of addition

in writing larger numbers. Using the 27 letters of the ancient

Greek alphabet, the highest possible number is 999. An

additional sign had to be used for larger numbers. The shape and the position of the additional sign is another difference in these two systems. There are also similarities and differences in indicating fractions in the two systems, however, this discussion is outside the scope of this book.

2. Anania Shirakatsi's Methodology

2.1. Net Effect of Anania Shirakatsi's Methodology

My research is based on only one part of the survived manuscript *Tvabanutiun* (Arithmetic), addition, the fundamental part of arithmetic, and the basics of the structure of his methodology. The manuscript contains interesting, detailed explanations of his methods that have been used for centuries in Armenia. The net effect of his teachings has resulted in the acquisition of a high degree knowledge in arithmetic by children, putting a strong foundation for subsequent achievements in scientific research in their mature age. As I will show, and explain my opinion that Anania Shirakatsi's teaching methodology can be as effective in the contemporary diverse classroom as it has been since the 7th century.

2.2. Anania Shirakatsi's Tables of Addition

In this book I limit my discussion only to addition. In ancient Armenian literature addition was referred to as *endunelutiun*, (receiving) in the sense that one number receives another number for being added.

Shirakatsi's tables of addition consist of four groups with nine tablets in each. In the manuscript all four groups are placed on a single page. Each group is placed in one column (see **Table A**). Please note that the table does not include the symbols + and =. The symbol + indicating addition was used for the first time in the 15th century, and the symbol =

indicating equality was first used in the 16th century[9]. To show the concept of tables more clearly, I have also shown Shirakatsi's tables in Hindu numerals instead of Armenian letters (see **Table B**). Notice that in **Table A** the first addend in each tablet is the same and it progresses downward in the alphabetical order. (We can check the same in **Table B**) The first addend in each tablet repeats, and it progresses downward from 1 to 9 in the first column, from 10 to 90 in the second column, etc. I will expand more on the significance of this setup after I describe the first group of tablets.

The first group is in the first column of the table. It shows the addition of single digit numbers. Note that in this group 2+1=3 is missing, as are 3+2, 4+3, etc. All tablets start with the addition of the number and itself. Then the first addend repeats, and the second progresses to 9.

---

[9] Ball 1901, p. 201

Why so? What can this tablet teach other than addition? The child understands that 1+2 is the same as 2+1, that addition is about putting two quantities together, regardless of order. In other words, it teaches that addition is augmentation, and that the order of addends is unimportant, that is, it teaches the commutative property of addition, and how to use it. Is it essential for a child to learn it? Absolutely! With this method the child understands the meaning of addition. The commutative property of addition is a very important discovery and a source of knowledge for the child. It helps to perform additions of numbers faster and intuitively. It becomes second nature, and stays in child's mind.

2.2.1. First Column: Addition of Single Digit Numbers

The first tablet of the first column teaches the addition of 1 to numbers from 1 through 9. The answers in this tablet, 2 through 10, are showing how numbers grow from 1 to 10. 1+1 is 2, 1+2 is 3, etc. The child understands that to count is the same as to keep adding 1 to the preceding number. The child learns that 1+3 is 4, and 1+4 is 5, that is, twice adding 1 to 3 we get 5, therefore, 5 is greater than 3, that 5 is 2 ones away from 3, and that the difference of 5 and 3 is 2. This also creates a good foundation for understanding subtraction. This is an important knowledge that will forever stay with the child.

Tablet two shows the addition of 2 to numbers 2 through 9. This tablet is one line shorter than the previous one. The third tablet shows the addition of 3 to numbers 3 through 9. Notice that this tablet, too, is one line shorter than the previous tablet. Each following tablet is shorter by one line. Finally, the last tablet shows the addition of 9 to 9 and consists of only one line. This, also, makes the child to stop and think. The student, looking at the tablets of the first column will see that the total of any two numbers in one tablet can also be found in other tablets. For example, three consecutive tablets show 1+5=6,

1 + 1 = 2	10 + 10 = 20	100 + 100 = 200	1000 + 1000 = 2000
1 + 2 = 3	10 + 20 = 30	100 + 200 = 300	1000 + 2000 = 3000
1 + 3 = 4	10 + 30 = 40	100 + 300 = 400	1000 + 3000 = 4000
1 + 4 = 5	10 + 40 = 50	100 + 400 = 500	1000 + 4000 = 5000
1 + 5 = 6	10 + 50 = 60	100 + 500 = 600	1000 + 5000 = 6000
1 + 6 = 7	10 + 60 = 70	100 + 600 = 700	1000 + 6000 = 7000
1 + 7 = 8	10 + 70 = 80	100 + 700 = 800	1000 + 7000 = 8000
1 + 8 = 9	10 + 80 = 90	100 + 800 = 900	1000 + 8000 = 9000
1 + 9 = 10	10 + 90 = 100	100 + 900 = 1000	1000 + 9000 = 10000
2 + 2 = 4	20 + 20 = 40	200 + 200 = 400	2000 + 2000 = 4000
2 + 3 = 5	20 + 30 = 50	200 + 300 = 500	2000 + 3000 = 5000
2 + 4 = 6	20 + 40 = 60	200 + 400 = 600	2000 + 4000 = 6000
2 + 5 = 7	20 + 50 = 70	200 + 500 = 700	2000 + 5000 = 7000
2 + 6 = 8	20 + 60 = 80	200 + 600 = 800	2000 + 6000 = 8000
2 + 7 = 9	20 + 70 = 90	200 + 700 = 900	2000 + 7000 = 9000
2 + 8 = 10	20 + 80 = 100	200 + 800 = 1000	2000 + 8000 = 10000
2 + 9 = 11	20 + 90 = 110	200 + 900 = 1100	2000 + 9000 = 11000
3 + 3 = 6	30 + 30 = 60	300 + 300 = 600	3000 + 3000 = 6000
3 + 4 = 7	30 + 40 = 70	300 + 400 = 700	3000 + 4000 = 7000
3 + 5 = 8	30 + 50 = 80	300 + 500 = 800	3000 + 5000 = 8000
3 + 6 = 9	30 + 60 = 90	300 + 600 = 900	3000 + 6000 = 9000
3 + 7 = 10	30 + 70 = 100	300 + 700 = 1000	3000 + 7000 = 10000
3 + 8 = 11	30 + 80 = 110	300 + 800 = 1100	3000 + 8000 = 11000
3 + 9 = 12	30 + 90 = 120	300 + 900 = 1200	3000 + 9000 = 12000
4 + 4 = 8	40 + 40 = 80	400 + 400 = 800	4000 + 4000 = 8000
4 + 5 = 9	40 + 50 = 90	400 + 500 = 900	4000 + 5000 = 9000
4 + 6 = 10	40 + 60 = 100	400 + 600 = 1000	4000 + 6000 = 10000
4 + 7 = 11	40 + 70 = 110	400 + 700 = 1100	4000 + 7000 = 11000
4 + 8 = 12	40 + 80 = 120	400 + 800 = 1200	4000 + 8000 = 12000
4 + 9 = 13	40 + 90 = 130	400 + 900 = 1300	4000 + 9000 = 13000
5 + 5 = 10	50 + 50 = 100	500 + 500 = 1000	5000 + 5000 = 10000
5 + 6 = 11	50 + 60 = 110	500 + 600 = 1100	5000 + 6000 = 11000
5 + 7 = 12	50 + 70 = 120	500 + 700 = 1200	5000 + 7000 = 12000
5 + 8 = 13	50 + 80 = 130	500 + 800 = 1300	5000 + 8000 = 13000
5 + 9 = 14	50 + 90 = 140	500 + 900 = 1400	5000 + 9000 = 14000
6 + 6 = 12	60 + 60 = 120	600 + 600 = 1200	6000 + 6000 = 12000
6 + 7 = 13	60 + 70 = 130	600 + 700 = 1300	6000 + 7000 = 13000
6 + 8 = 14	60 + 80 = 140	600 + 800 = 1400	6000 + 8000 = 14000
6 + 9 = 15	60 + 90 = 150	600 + 900 = 1500	6000 + 9000 = 15000
7 + 7 = 14	70 + 70 = 140	700 + 700 = 1400	7000 + 7000 = 14000
7 + 8 = 15	70 + 80 = 150	700 + 800 = 1500	7000 + 8000 = 15000
7 + 9 = 16	70 + 90 = 160	700 + 900 = 1600	7000 + 9000 = 16000
8 + 8 = 16	80 + 80 = 160	800 + 800 = 1600	8000 + 8000 = 16000
8 + 9 = 17	80 + 90 = 170	800 + 900 = 1700	8000 + 9000 = 17000
9 + 9 = 18	90 + 90 = 180	900 + 900 = 1800	9000 + 9000 = 18000

Table B: Shirakatsi's tables of addition shown in Hindu numerals.

2+4=6 and 3+3=6. Thus, the student will conclude that 6 is the addition of 1 and 5, or 2 and 4, or 3 and 3. This teaches the value of 6, that 6 is made of 1 and 5, or 2 and 4, or 3 and 3. This knowledge later will help to add numbers without using calculator faster. For example, to perform 8+5 the child, knowing that 8+2=10 and 5=2+3 will be able to perform 8+5=8+2+3=10+3=13.

As I mentioned above, the first tablet teaches the child to count from 1 to 10. In the second tablet the sums are in the same sequence, and 11 follows 10. Examination of subsequent tablets will teach the child continuity in counting.

2.2.2. Second Column: Addition of Tens

The second column shows addition of tens. The first tablet in column two helps the child to count in tens: 10, 20, 30, etc. They will connect counting in tens with the order of letters in the alphabet. Having the first and the second columns side by side will help the child to compare them and find similarities and differences. The child who has already mastered the addition of single digit numbers and has already worked out a method to ease the addition of single digit numbers will use the same method to add tens.

Similarly, **Table B** can be effectively used to teach addition. Having the first and the second columns side by side helps the child to compare them, and to find similarities and differences. Seeing 2+3=5 and 20+30=50 side by side, the child notices the similarities (numbers 2, 3, 5), and will notice the difference, 0's after these digits. The child will learn that 20+30=50, meaning 2 tens and 3 tens, make 5 tens or 50. This addition builds up the knowledge presented in the first column, thus teaching the idea of the place value.

In adding 20 to 30, the present-day student who is taught to perform addition using the "Addition Algorithm" will invariably, by rote write 20 and 30 one under the another, and will write: "0+0=0, 2+3=5, therefore, the answer is 50."

2.2.3. Third and Fourth Columns: Addition of Hundreds and Thousands

Next is the third column, the addition of hundreds. Building on the knowledge gained from the first and second columns, the child develops better knowledge of the place value and learns the addition of hundreds.

The fourth column, the column of addition of thousands teaches the pattern of addition of ones, tens, hundreds, and thousands. This presents the idea of progression and continuity, the idea of infinity.

Likewise, in calculations with thousands, to add 2000 to 3000 the present-day student who is taught to perform addition using the "Addition Algorithm," again by rote will invariably write 2000 and 3000 one under the another, and will write: "0+0=0, 0+0=0, 0+0=0, 2+3=5, therefore, the answer is 5000." The student fails to see the connection between 2+3, 20+30, 200+300, and 2000+3000, which in turn prevents noticing progression and continuity.

3. Conclusion

Initially, one may argue that this is a very slow pace in learning, and children nowadays are smarter than they were in the 7th century, that children in the 21st century have calculators and computers, and consequently they do not need to learn how to add. What if, accidentally, he/she pushes the wrong button, or types one 0 more? Will the person notice that the answer he/she arrived at is ten times larger? In my experience, they often do not notice it. Through learning to add using Shirakatsi's methods, the child develops an ability to identify how large or small the presented numbers are, and learns to approximate the answer in the head, while performing calculations on a calculator. Therefore, he/she will definetly catch the error occurred because of typing one 0 more or less, or pushing the wrong button of the calculator.

Strong foundation guarantees success in learning. Other than teaching the child how to add, these methods teaches the child to understand the meaning of addition, and teaches good understanding about the value of numbers. These methods teach thinking and encourage reasoning. From an early age, thess methods teach the child to separate issues, to analyze, to ask questions and to look for solutions. As a result, logical thinking and reasoning become second nature for the child. This is due to, in Shirakatsi's own words, that he starts his teaching from its fundamental level, and presents the concepts in a most simplified manner for thorough understanding.

I have reviewed a number of elementary level textbooks of arithmetic where instead of teaching addition with a systematic explanation, yes, maybe slow pace, they jump into the addition of arbitrary pairs of numbers. It is the previously mentioned "Addition Algorithm," an unknown and hard to pronounce words for the child. It alienates the child and creates unnecessary distraction. Invariably, this method resorts to counting bunches of sticks haphazardly, that is, the child counts 5 sticks, then counts 3 sticks and puts them next to the 5 sticks and counts all sticks again to come up to a total of 8. The child, not connect 5+3 and 3+5, will count 3 sticks, then will count 5 sticks and will put them next to the 3 sticks, and then count all the sticks again for a total of 8, getting confused all along. Boring? Yes! I have seen how some of my grown-up students count on their fingers. Teaching addition through repetitive drills teaches to memorize that 5+3 is 8, 4+7 is 11, 3+5 is 8, but does not teach why. This does not encourage the child to think, to understand, to look for solutions. Drills are very boring for the child.

There are "algorithms" for everything. Students associate

the word algorithm with 'steps' that they have to memorize. This eliminates the necessity to understand, and learning mathematics becomes exercising memorization of 'steps.' Furthermore, students have to memorize all possible "problem and solution algorithm" combinations. "Too much common sense is missing today from our educational system." states Dr. Joseph Casbarro, the author of several books on teaching[10]. Learning mathematics through memorization is not a long lasting learning. Mathematics builds up on previously gained knowledge, and a weak foundation makes learning mathematics impossible. All these distractions, boredom, and frustration make learning math distasteful, hard, if not impossible. Despite my own encouragement in understanding, versus memorizing, my grown-up students habitually try to match the problem with given 'steps.' The use of briefly memorized 'steps' eases problem solving, at the expense of understanding the problem. (After forgetting the 'steps,' to

---

[10] Casbarro 2003, p. XVII

salve the same problem usually becomes challenging for the student.) The question "Why do we do these 'steps'?" has been replaced by the question "What are the 'steps'?" Furthermore, students often do not use their common sense to check if the answer of the problem makes any sense at all. Once I assigned to my students to calculate how much gas was needed to drive one mile, if with six gallons one could drive 150 miles. One student immediately responded, "900 gallons," applying multiplication instead of division.

ABOUT THE AUTHOR

Gohar Marikyan is a Professor of Mathematics and Computer Science at the State University of New York Empire State College in Manhattan. She holds a Ph.D. degree in Mathematics and MS degree in Computer Science. Dr. Marikyan continues to actively pursue her research in mathematical logic and computer science. She also is interested in the history of math education. The results of her research she presents in domestic and international conferences and publishes in a number of scientific peer reviewed journals.

Gohar Marikyan is a recipient of the Susan H. Turben Award for Excellence in Scholarship.

Born, brought up, and educated in Armenia, which was then part of the Soviet Union, Dr. Marikyan had to study the history and ideology of the Soviet Union, and was an eyewitness to its workings and its collapse. After the collapse Dr. Marikyan was involved firsthand in the development and the operation of the very first nongovernmental organization in the Soviet Union – the Gtutiun Armenian Charitable Union – as its information systems department head, liaison officer, and publicity director. She won a grant by the United States Information Agency's Regional Scholar Exchange Program and Freedom Support Act Fellowship. Dr. Marikyan also is a certified Oracle database administrator, and has a Master of Sience degree in teaching religion from the Armenian Apostolic Church.

Gohar resides in New York City, in Manhattan. She is currently working on a novel titled *I Wish I could Touch you Again* from the series *Life is Stranger than Fiction*. She also continues her research on Anania Shirakatsi's 7th century manuscript that will culminate to *Anania Shirakatsi's Tvabanutiun: World's Oldest Manuscript On Arithmetic, Part 2* book. The third project is a colorful book on Armenian dolls (see www.aaadolls.com).

For more information on new publications and to ask questions please visit Gohar Marikyan's website at www.GoharMarikyan.com.

OTHER PUBLICATIONS

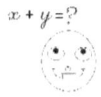

Love, Death, and Roses: Life is Stranger than Fiction, by
Gohar Marikyan, Ph.D.

This book is dedicated to Lena, a strong and positive woman. Her life story sounds like a soap opera that we would very much want to be true. We also like to find out if her son Yura really exists. Rest in peace, Lena jan.

The book can be found at www.amazon.com. Is available also in Kindle version.

Zakooda Shendi's Legacy
Discovered by Aramazd Babakhanian
Designed by Vardan Stamboltsyan
Preface by Gohar Marikyan, Ph.D.

The origin of the epigrams published in this book is unknown. The epigrams are unique and provoke us to think about the life, love and relationships at unknown times in an unknown country. They were handed over to me by Mr. Aramazd Babakhanian. His last will was to have them published.

Not a letter, punctuation or formatting of the epigrams has been changed. To some of the epigrams the first line is added as a title. They include words that might provide clues to their provenance.

The book can be found at www.amazon.com. Is available also in Kindle version.

Index

7th century, 7, 10, 15, 17, 19, 20, 31, 39
Abrahamyan, 12, 53
adding, 12, 13, 34, 37
addition, 8, 12, 21, 22, 30, 31, 32, 33, 34, 36, 37, 38, 40
algorithm, 21, 22, 42
alphabet, 13, 15, 24, 28, 36, 37
analyze, 40
Anania. *See* Shirakatsi
Anania Shirakatsi. See Shirakatsi
ancient, 16, 17, 24, 28, 31
Arabic, 14, 16, 17, 32
arithmetic, 7, 8, 10, 12, 20, 21, 22, 30
Armenia, 7, 10, 14, 16, 17, 23, 24, 30
Armenian language, 14, 15
Ashkharhatsuyts, 17
Assyrian, 16, 17
autobiography, 17
Bible. See Holy Bible
boredom, 42
boring, 21, 41
byur, 27
Catholicos, 13, 15
Chapagitutiun, 17
children, 7, 8, 10, 20, 21, 22, 30, 39
Chronicon, 17
civilization, 9
classroom, 8, 31

common sense, 42
commutative property, 33
continuity, 36, 38
cosmologist, 12
Cosmology, 17
count, 12, 34, 36, 41
counting, 9, 36, 41
distraction, 41
division, 12, 43
drills, 41
educator, 10
effective, 7, 31
Ethiopian, 16, 17
foundation, 7, 22, 30, 34, 40, 42
fractions, 29
geographer, 12
Geography, 17
Golden Age, 14
Greek, 12, 16, 17, 23
Grigor Magistros, 15
Hacikyan, 12, 24
Hebrew, 16, 17
Hewsen, 7, 12, 54
Hindu-Arabic, 24
historian, 12, 23
history, 9, 10, 14
Holy Bible, 18
hundreds, 25, 37, 38
Indian, 16, 17
infinity, 38
Inqnakensagrutiun, 17
Iranian, 16, 17
languages, 16, 17
lasting, 33, 35, 42
Latin, 16, 17
letters, 14, 15, 24, 27, 32, 36
literary, 14
literature, 18, 31
logical thinking, 40
manuscript, 7, 10, 16, 17, 20, 22, 30, 31
Mashtots, See Mesrop Mashtots
Matenadaran, 16, 17
mathematician, 10, 11, 15, 17
Mathematics, 9, 24, 42
Measures, 17

memorize, 21, 22, 41, 42
Mesrop Mashtots, 14, 15
Meteorology, 17
methodology, 7, 8, 10, 30
methods, 7, 9, 20, 30
multiplication, 12, 43
non-negative numbers, 12
numbering system, 14, 24
numeral system, 27, 28
numerals, 15, 24, 32
Odyerevuitabanutiun, 17
philosopher, 12
philosophical, 16
place value, 25, 37, 38
positional system, 25
progression, 38
reasoning, 40
religious, 14
research, 7, 8, 30
rote, 21, 22, 37, 38
Sahak Partev, 13, 15
Saint Mesrop. See Mesrop Mashtots
scholar, 14, 15, 16, 19
scholars, 15, 17
science, 9, 13, 14, 21
scientific, 7, 10, 16, 30
scientific works, 10
Scriptures, 18
second nature, 33, 40
Shirak, 18
Shirakatsi, 7, 8, 10, 15, 17, 18, 19, 20, 21, 22, 23, 30, 31, 40
single digit numbers, 32, 36
steps, 42
subtraction, 12, 34
system, 24
teaching, 7, 8, 9, 10, 20, 21, 22, 31, 40, 42
teaching mathematics, 9
ten fingers, 9
ten-based system, 9
tens, 25
thinking, 10, 40
thousands, 25
Tiezeragitutiun, 17
Tomaragitutiun, 17

Tvabanutiun, 7, 16, 17, 20, 30
understand, 21, 40, 41, 42
Vramshapouh, 13, 15
writer, 11
writings, 15
Yerevan, 16, 17
zero, 25

References

1. Abrahamyan, A., 1944, Anania Širakac'u

   Matenadrut'iwn [The Works of Ananias of Širak],

   Yerevan (in Armenian).

2. Ball, W. W. R., 1901, A Short Account of the History of

   Mathematics, New York: Macmillan.

3. Casbarro, J., 2003, Test Anxiety & What You Can Do

   About It: A Practical Guide for Teachers, Parents, and

   Kids, Port Chester, NY: Dude Publishing/ National

   Professional Resources, Inc.

4. Hacikyan, A. J., Basmajian, G., Franchuk, E. S., Ouzounian, N., 2002, The Heritage of Armenian Literature, Detroit: Wayne State University Press, vol. II.

5. Hewsen, Robert H., 1968, "Science in Seventh-Century Armenia: Ananias of Širak", in Isis, Vol. 59, No. 1, (Spring, 1968), pp. 32-45.

6. Rosenqvist, J. O., 2005, "Byzantine Trebizond: a Provincial Literary Landscape", in Byzantino-Nordica 2004. Papers presented at the international symposium of Byzantine studies held on 7-11 May 2004 in Tartu, Estonia, P. Ivo & J. Päll (eds.), Tartu, Estonia, Acta Societatis Morgensternianae, pp. 29-51.

7. Shirakatsi, A., 7th Century, Tvabanutiun (Arithmetic), Manuscript.

8. Shirakatsi, A., 7th Century, Autobiography, Manuscript.

9. Aghayan, E.B. (Editor), 1979, Anania Shirakatsi, Selected Works, Yerevan: Sovetakan Grokh.